THE STORY OF
SIERRA LEONE

BY

A. P. KUP

CAMBRIDGE

AT THE UNIVERSITY PRESS

1964

PUBLISHED BY
THE SYNDICS OF THE CAMBRIDGE UNIVERSITY PRESS

Bentley House, 200 Euston Road, London, N.W. 1
American Branch: 32 East 57th Street, New York 22, N.Y.
West African Office: P.O. Box 33, Ibadan, Nigeria

©

CAMBRIDGE UNIVERSITY PRESS

1964

Printed in Great Britain at the University Printing House, Cambridge
(Brooke Crutchley, University Printer)

Contents

List of Plates

between pp. 32 and 33

1   An ivory salt-cellar.

2 (*a*)   Suri, King of Sherbro, 1758.

  (*b*)   Governor Rowe addressing chiefs.

3 (*a*)   The *Black Joke* captures the *Almirante*.

  (*b*)   View of Freetown in the 1870's.

4   Reclining figure carved in soap-stone.

PLATES 1 and 4 are reproduced by courtesy of the Trustees of the British Museum, 2(*a*) by courtesy of Messrs Routledge and Kegan Paul, 2(*b*) by courtesy of *The Illustrated London News*, and 3(*a*) and (*b*) by courtesy of the Trustees of the National Maritime Museum.

Introduction

HISTORY is the word we use to mean what has happened in the past to people, or things. One could have a history of one person (although this would be difficult because everyone lives with, or near, someone else) or even the history of one thing. But usually histories are written about groups of people *and* things. This history, *The Story of Sierra Leone*, is about everyone in Sierra Leone as well as about such things as trade, money, house-building, food, government and war. Of course it does not mention everyone, or everything, otherwise it would take longer to write than the events took to happen.

Just as a man who builds a house chooses his materials from different places—mud from the river, cement from the store and wood from the forest—so does a writer of history.

When people cannot write they tell stories to one another and the more famous of these are remembered for hundreds of years. We have one of them in the first chapter; it was first written down in his own language by a Frenchman in the 1930's, when he heard it from the elders of the tribe. There are many stories of this kind which have never been written down, but they are still part of history and they are very important indeed from our point of view.

There is another way for a historian to gather material. When people die they leave things behind

5

them—cooking pots, axe heads, and even rubbish heaps. All these and many other tools and weapons help the historian to build up his picture. You may see a few of these in the Sierra Leone Society's museum in Freetown; all of them have been made and used in Sierra Leone by people long since dead. From them we can tell how people used to work on their farms, how they made beads, or what weapons they used to hunt wild animals.

Finally, there are written sources of history: old letters, diaries, account books, treaties, government reports. Sometimes carvings, paintings and photographs exist which tell us what life was like, or what people wore in early times. Written records of this country begin about A.D. 1500, but they are not common until about 1800; photographs began in the 1800's.

Just as a mason or a carpenter today does his best to make a straight or 'true' line from his materials, so must an historian do the same. Not all sources are useful; some are not truthful and others are not complete. Some valuable information may be lacking altogether. Others show only what one person thinks, and tell us nothing about the other side of the question. An historian's task is to make a true story as best he can. Like a bird on a farm which pecks up what it can find after the harvest, the historian picks up all that he can from what is left of the past.

Thus our chapter 2 (except for the boy Barki who has been invented) is drawn from several early stories written about this country by Portuguese and English traders. No single account tells very much, but by putting them all together a fuller story can be built up.

6

Chapter 3 is taken largely from a report of a Portuguese Jesuit missionary called the Reverend Balthasar Barreira who came to Sierra Leone in 1605 and who stayed several years.

Chapter 4 comes from many different sources; the man who wore a rabbit's skin next to his stomach was Jean Barbot, a Frenchman who wrote a long book about the West coast. He copied much of his story from a Dutch book by O. Dapper, a doctor who wrote in the 1660's.

The European trader in chapter 5, who lived on the River Jong, was an Irishman called Nicholas Owen, whose hobby was making decorations out of seashells. In his diary, amongst other valuable facts, he has left us a picture of Suri, King of Sherbro, which he drew in 1758. There were, of course, no cameras then and this is our first named picture of any ruler in Sierra Leone.

Much of the story of James Somerset in chapter 6 comes from a book *The Memoirs of Granville Sharp*, published in 1820. The arrival of the Settlers is told by a Swedish business man, Wadström, who was interested at that time in founding a colony in Africa for religious reasons.

The description of Freetown in 1793 is from a diary—or log-book as it is called—of an English ship commanded by Captain Samuel Gamble, a slaver. This, like other accounts used in writing this book, has never been printed; it is now in a museum in London, just as Gamble wrote it when his ship lay at anchor off Freetown so many years ago.

7

Many details in chapter 7 come from the letters written home by a Portuguese official who was a judge in the court which condemned slaving ships after they had been captured by the British Navy and brought to Freetown, where the slaves were set at liberty. This court was called the Court of Mixed Commission, because it had as its judges people of different or mixed nationalities. Other details are from the books or speeches of two famous Sierra Leoneans, E.W. Blyden and Sir Samuel Lewis.

Chapter 8 is drawn mainly from official reports.

Chapter 9 is a history of certain ideas, and ideas are often more important than people or things. Those who have new ideas produce new machines or help the farms to grow more food; they create new forms of government. It was a new idea, for example—that of finding a sea way to India from Europe—which brought Europeans to the West Coast of Africa, and so in time produced European colonies here. It is another new idea which is now leading these African peoples to independence again.

One should also know that the name Sierra Leone was first used by the Portuguese, and that it meant only the hills round what is now Freetown. As happened with the names of some other modern states in Africa, it was only during the late nineteenth century that the words Sierra Leone came to mean our country as it is at present.

Map 1 Sierra Leone in her world setting

9

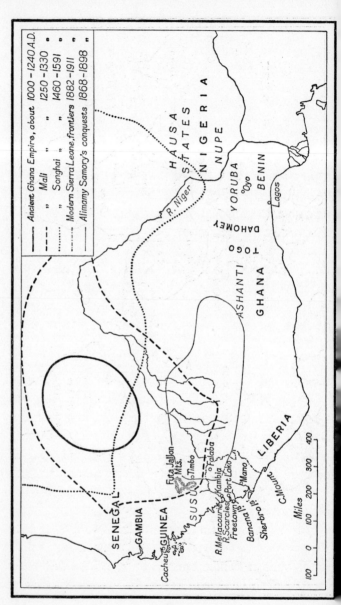

Ancient Ghana Empire, about	1000 – 1240 A.D.
" Mali "	1250 – 1330
" Songhai "	1460 – 1591
Modern Sierra Leone, frontiers	1882 – 1911
Alimamy Samory's conquests	1868 – 1898

1. *The First People in Sierra Leone*

THE EVENTS in the story we are to hear now took place long before the first Settlers came in 1787; before even the first white men came to this country in the 1460's.

Where Freetown stands today there then were only three or four villages, and many elephants walked through the bush. Parrots flew across the river to feed in the daytime. Hippopotamuses swam in the river and bush-cows grazed in the spaces in the thick forest.

Important men wore clothes very like those of the Fulah of today; these were brought down from the north by the Susu, who were paid not in money, but in salt which was taken from the sea. Women wore beads made of bone and a cloth round their hips.

It is not easy for us, after more than five hundred years, to find out much about these early people, because they did not know how to write and so they have left us no writing about themselves. But a few stories have been handed down by word of mouth, and also some of their tools and cooking pots have been found—sometimes by diamond diggers—buried deep in the ground or in caves. Their tools were often made from pieces of stone which they shaped into axe heads or hoes or weights for their fishing lines.

One of the first stories we have of these people tells us how the Susu moved into the country just north of modern Sierra Leone. This was about the year A.D 1400. By then people had learnt many things, such as how to make iron blades for their spears and how to grow new kinds of food. But food was not always easy to get and we shall see that the Susu story says that they moved towards the west—towards the setting sun, as they said. They were following the herds of elephants whose meat they liked very much.

Here is the story:

The Susu were hunters. They went towards the west with their women and children, moving slowly like ants. They lived by hunting and by gathering roots and wild berries. They were armed with arrows and had dogs as fierce as leopards. They also fished in the rivers.

When they rested, their leader Domin Konteh set up his family near him in huts made of wood and fibre. The men hunted, the women made clothing out of the skins of wild animals.

One day there was trouble in the north-west. The Susu met some tall black strangers from the north who wore no clothes. These people were the Baga. However, after some fighting the Baga settled peacefully in lands all round the Susu.

When Domin Konteh died, his son, Manga Kombeh Balla, became the leader of the Susu. About this time the first Portuguese traders arrived, but they did not settle; they lived in their ships on the sea.

This story tells us about the Susu and the Baga who are relations of the Temne; elsewhere it tells us of the Fulah and Yalunka.

Soon the Portuguese came to settle and they began to write stories about Sierra Leone. From these we learn of other people in this country; the Sherbro-Bullom, the Temne, the Limba, the Loko, the Gola and the Krim. All these people were amongst the first in Sierra Leone.

The early Portuguese visitors could not speak any African languages; but fresh water is important to any sailor or fisherman on the sea and they soon learnt the words for this. They also wanted to buy gold and they learnt African words for that too. They found two words for water and two for gold—one was a Sherbro word and the other was a Temne one. We know from this that the Sherbro and Temne were living on the coast when the Portuguese came.

We shall hear about the Limba and the Loko in the next chapter, but about the Gola and the Krim in the south of Sierra Leone we know very little before 1500. The Portuguese, who were the first white men to come to Sierra Leone, spoke of a people to whom they gave a name like, but not exactly the same as, the 'Gola'.

They knew too of a town in the south which lay some way up-river from the sea; it was a famous town with many houses. Here the Portuguese traded in a kind of tin called pewter. Much rice grew there, but the place was unhealthy for Europeans and many died because there was so much fever in the district. When the Portuguese tried to spell this town they wrote *Quimanora*—for Krim Mano, that is, Mano in Krim country.

Unlike the Susu, the Krim have no history of having moved into their country from anywhere else, so they must have been in Sierra Leone for some years before the Portuguese came to this country.

2. *The Mani invade Sierra Leone*

EARLY one morning in the year 1564 a young boy called Barki was drawing water from the river which ran near his town close to Sherbro Island. As he stood upright he looked through the palm trees beneath which the goats were feeding. Down river he saw three or four large canoes with fierce strangers in them paddling against the stream. Calling in fear to his friends he ran quickly back to the town where soon his shouts brought the elders to him.

The alarm was spread; Barki, the women, and other children were sent to hide in the swamp whilst the warriors took up their arms and ran to a place where the road from the river to the town was steep and narrow. Each man carried a bow, a spear and a shield made from tough elephant skin; and they had long arrows, made from reeds, which were poisoned. Here they hid and waited in silence for the strangers.

Barki with his friends, safe in the swamp, could hear the shouts and the drums as the two sides began to fight.

Then there was silence. Who had won? Was it safe to go home? Why had the strangers come? Who were they? Many questions like this were asked but no one knew the answers until, hearing a friendly shout, Barki and the others all ran back from the swamp to their town.

What a sight! Children and mothers looked anxiously for their fathers and husbands, some crying aloud when they saw their wounds, others glad to see them safe. But it had been a great victory; prisoners had been taken and a big warrior, the leader, had been killed by the king of Barki's town.

Although they did not speak the same language, the prisoners had explained by signs that the great warrior's name was Masariko, that they called themselves Mani, and that they had come up the sea-shore from the south. Barki noticed that they did not file their teeth as some of his elder brothers and uncles did, and that they had on their bodies few marks such as were made at certain ceremonies in his town. Their bows and arrows were smaller too. Barki noticed that they looked in wonder at the corn, rice and roots which grew beside the town; it was as if they had never seen any before.

That night there was feasting and dancing, but Barki saw that many of the elders looked worried whilst they drank their palm wine. In a few days he learnt why: these Mani strangers were part of a huge army, much too large to be attacked with success by the people of the scattered towns in Sherbroland. So Sherbro kings had agreed to ask for peace.

There was much talk about the strangers; some said they had conquered the people to the south (the Gola and the Krim). Others said it was not true, the southern peoples had hidden in the swamps so that the Mani had passed them by. Others again said that the Mani had not dared to march towards the north because they feared the warriors there who lived in strong towns made of stone (these were the Limba and the Yalunka).

Soon proud Mani kings and princes were living in the best Sherbro houses. They had with them a few Portuguese, who had guns, and fearsome warriors whom, it was said, they had trained to eat human flesh. Barki had never heard of such a thing before and he was very much afraid. Also, although the Mani were good sailors and handled their canoes well, they knew nothing of goats or cattle, nor did they understand farming. They said all Sherbro warriors must join their army, but because Barki was too young to fight he was made to plant rice and corn and to gather wild fruits for his new masters.

Presently the Mani leaders began to advance again. They sent messengers to the Temne with two kinds of gifts; one of cloth and the other of weapons. To accept the first was to accept Mani friendship; to take the weapons was a sign that one wanted to fight.

The Temne took the weapons, so the Mani marched against them. Many Temne fled, some even running to Portuguese ships which lay along the coast, because they preferred being slaves in the West Indies to being eaten by the Mani soldiers. These Temne called the Mani *Sumba*, which means 'those who cause trouble'. Other Temne fought bravely. Later Suri, the Temne king of the Sierra Leone peninsula, surrounded a Mani camp on the banks of the Sierra Leone River.

The Mani, unlike the people of this country at that time, had strengthened the outside of their camp with big stakes and trees, and this caused much trouble to Suri who had not seen this kind of defence before. As it happened, some English ships had arrived in the river and, with the help of 90 sailors, Suri attacked the enemy. This was in January 1568.

Twenty Englishmen were wounded; some fell into hunters' traps—holes in the ground covered with grass and full of sharp stakes. In the end hot cannon-balls were fired into the camp and in the confusion Suri was able to break down the walls of logs. He won the fight, but his son was killed and two Mani leaders, Sarina and Seterama, escaped.

At length the Temne also were conquered, and they too made an alliance with the Mani. Sometimes the fighting was so fierce that there were no more arrows left. The Mani with their smaller bows could pick up and shoot back the longer Temne arrows, but the Temne bows were too long for the small Mani arrows, so that when the Temne had fired all their own arrows they had to retreat.

The Loko also were overcome and a Mani warrior, Bai Farma, became their king. But he did not stop fighting, as we shall see.

Soon the Mani, Sherbro and Temne advanced north upon the Susu who had come to meet them near the River Scarcies. Amongst the Sherbro was Barki, now a man and able to rule men under him. But the Susu were clever; they had sent out parties who pretended to be feasting. Cattle were cut up, the meat was poisoned and, when the Mani came, the Susu ran off as if in fear. They left the pots boiling on the fires. The Mani, suspecting nothing, fell hungrily upon the food. Many died. Alas! Barki and his warriors were amongst the dead.

17

Next the Susu borrowed seven Fulah horsemen who were fine warriors, and when the battle began the Mani, who had not yet fought against horses, were so frightened that they quickly fled.

They retreated behind the River Sierra Leone and thus, after seven years' fighting, there was peace again in this country.

3. *King Tura and King Farma*

FARMA fought well for the Mani during their invasion—so well that his leader, King Tura, King of all the Mani, made him the first Mani ruler of the Loko. His country stretched inland for many miles and was, by 1600, the largest in those parts. An English sailor who met him in the Sierra Leone River in 1582 said that Farma had two or three hundred wives and that his riches lay in slaves and ivory. For many years he fought with the Limba to make his kingdom larger still. He died in 1605 after he had been poisoned because he was so harsh. It was said that he was 130 years old when he died. He left a sister, Mabora, who must, like her brother, have been a very fierce person. When the army went to war she placed herself at the back, armed with a bow and arrow, with a stick in her hand which she used on any warrior who tried to run away.

Farma divided his kingdom between two of his sons Sangrafaré and Souga. Sangrafaré was given the Loko kingdom and, though he had been crowned by hi

18

people, an elder brother drove him out, so that he had to fight before he could get his kingdom back.

Souga was given the peninsula of Sierra Leone—more or less the area of the Western Area today—but he said he did not want it, and he acted instead as Speaker to another brother, Philip, who ruled in his place. Souga was well liked and all listened to what he said. He told the Kings and Headmen in the mountains south of the Sierra Leone river-mouth to give a piece of land to the Portuguese so that they could build a fort near the river. His brother Philip wrote to the King of Spain, who then ruled also over Portugal, giving permission for this to be done.

Philip is a European name; we do not know his African one. He and four of his children had been christened by a Catholic priest called the Reverend Balthasar Barreira, who was the first Christian missionary to live in this country for any length of time. Philip therefore was the first Christian king of Sierra Leone; he lived in a town near Kru Bay which was called Salvador. He spoke Portuguese well because many Portuguese then lived in Sierra Leone; indeed Salvador is a Portuguese name.

A church was built in this town and before it was finished Barreira said his first Mass there on 29 September 1605. Later a wooden cross, which Philip himself helped to carry, was placed near the beach where the ships came in.

Other churches, made of wood like the one in Salvador, were built in Tura's kingdom, which lay a little further inland than Philip's, near the Sierra Leone

River. Tura's son had seen Barreira celebrate his first Mass in Sierra Leone and he told his father about it.

King Tura wanted to see this new wonder and on Christmas Eve 1605 he came with many musicians to watch the Christmas service. The priest had covered the walls and floor of the church with branches; on the altar was a green cloth and the whole building was lit with bright torches and candles. When he came to the door, Tura was filled with wonder and kneeled down looking for a long time at the strange sight. After the service he said that he wanted churches in his country too.

Often words and names help to tell us where people come from.

Mande-speaking peoples used the word *Tura* to mean 'a bull'; and one of Tura's sons was called Jata. Now, Mari Jata (which means in English 'Prince Lion') was a famous warrior who defeated the old kingdom of Ghana and began the great Mandingo empire of Mali about A.D. 1240. He was also called Sundiata.

Mandingo is a word used to mean 'those who speak the Mande language'. In Mande the word *Mali* means 'where the king lives' and so can be used as the name of any head town, or capital, of any Mandingo state.

In fact, the words *Mande*, *Mali*, *Mani* and *Mende* are all closely related. They have been used at different times in history for related peoples.

Farma is also a Mandingo title, meaning 'king' or 'governor'. *Bai*, however, is a Temne title and it was given to King Farma by those Temnes he had conquered in Sierra Leone; this king is known today in Sierra Leone as Bai Farma.

These words help to tell us something about the history of the Mani. They were a Mande people, once part of the old kingdom of Mali. After travelling many years they entered Sierra Leone. They married into Sierra Leone families and very soon it was not easy to say who was or who was not the grandson of a Mani.

4. *The Early Europeans*

THE PORTUGUESE were the first Europeans to come to this country, and they were the first to give it a European name. Their leader, Pedro de Sintra, called it in his language *Serra Lyoa*, that is 'wild mountain'. Later this was changed to the English form Sierra Leone. Another Portuguese, Governor of Elmina Castle in Ghana from 1520 to 1522, wrote: 'Many people think that the name *Serra Lyoa* was given because lions were plentiful in those parts, but this is wrong. It is because Pedro de Sintra, when he saw the land looking so rough and wild, called it *Lyoa*. This is the only reason; he told me so himself.'

By 1600 many Europeans were trading in Sierra Leone. National companies, formed by groups of people who joined together to trade, had been set up. These had special rights of their own in Africa, and nobody else in Europe was allowed to buy or sell in their district. But this did not stop private traders from visiting or even living in the country; such men were called interlopers.

The white men did not come without being afraid of what they might find: the sun (so much hotter than in Europe), the strange animals and foods, the black skins of the people they met—all these filled them with terror. Strangers first came to West Africa either on a camel or a horse, like the Arabs, or in a sailing ship, like the Europeans. The winds and currents in the sea were difficult for such ships and often sailors died of thirst or hunger because they could not get home soon enough. The voyage out to Sierra Leone took about eight weeks, and the return journey against wind and current somewhat longer.

In 1583 an Englishman wrote: 'On the 24th April we came to the coast of Guinea. Here we found it so hot that all the drinking water began to smell very bad and the sailors had to hold their noses whilst they drank.' A few years later another Englishman advised his friends to take water casks made so that they did not leak in the hot sun when the wood became dry and to have a great store of bread and keep half of it for the voyage home. Salted beef and pork, he said, would last for two years and still be good to eat. Vinegar should be carried to clean the ship so as to keep away disease. There were no antiseptics as we know then today; there were no trained doctors either.

At the end of the seventeenth century, one European died on the coast about every ten days. Many strange plans were made in order to keep fit; one man wore a rabbit's skin over his stomach with the fur next to his own skin. No one knew until about 1900 that it was the mosquito which carried malaria, so no one used nets. But if there were diseases on the coast which were fatal to white men, Europeans also brought their own diseases which proved fatal to the Africans. Smallpox was very dangerous. About 1600 there was an outbreak of measles near Cape Mount, brought by Europeans from Sierra Leone, which killed most of the people there.

The Portuguese, the first Europeans in Sierra Leone, traded only from their ships; later they settled here, mainly at Port Loko. Later still, when they had given up Sierra Leone to the English, they traded from Cacheu up the River Scarcies, especially for cola nuts which, they said, were the best on all the coast. They never built a fort on the land given them by King Philip of Sierra Leone, because soon after they had begun to build they were attacked by a French ship which carried away all their food and materials and left them to starve.

The English followed the Portuguese; after them came the French and then the Dutch. The English had a house, or factory as it was called, in Sherbro by 1651. The French were more interested in the countries north of Sierra Leone. The Dutch, though they had a factory in the Gallinhas for a few years and a small settlement in the Sierra Leone River for a short time, settled mostly in the Gold Coast, as Ghana was then called.

These visitors brought to Sierra Leone goods for

trading, such as brass and pewter kettles, iron bars, cloth, guns and gun-powder, bugles, knives and beads. They took away Sherbro mats, redwood for dyeing, indigo, animals' skins, and sometimes slaves, though only a few came from Sierra Leone itself.

The nations who had fought for so long in Europe did not, of course, stop fighting when they came to Africa. The Portuguese murdered the whole crew of a French ship which had found its way to Port Loko, because they did not want anyone besides themselves to know of the place. Both the Dutch and the French attacked the English settlements on Tasso, Bunce and York Islands. Indeed this was why the English built their headquarters on these islands: they were easier to defend.

The Dutch Admiral de Ruiter, who took Tasso Island from the English in 1664, carved his name, as did some of his comrades, on a large stone near the stream where most visiting sailors drew their fresh water. This place in Kru Bay is now called King Jimmy. The stone is now buried in the sand. In early times the water there was famous in Europe as the best on the coast for many miles; often captains filled their casks from it before making the long crossing to the West Indies or to North and South America.

The fort at York Island has been washed away by the tides and river floods, but much of the buildings at Bunce may still be seen, although the slipways on the east side, where ships were built and repaired, have rotted away. The English first built a factory there in the 1670's, but they still used the drinking water from

24

Tasso Island and they also grew most of their vegetables there. However, the first buildings on Bunce Island were destroyed and those which can be seen now date mostly from the eighteenth century. Together with the Ruiter stone at King Jimmy and a tombstone on the Banana Islands put up to the memory of Captain Reid, Royal Navy, who died there in 1713, they are our only relics of early European settlement.

5. *The Coming of the Mende*

No WRITER mentions the Mende by name until the seventeenth century. Of course, this does not mean that they did not exist, but that they lived further inland than they do now, which meant that Europeans, who lived near the sea, did not meet them. As we have seen, the Mani and the Mende were close relatives, but the Mani were few in number when they entered Sierra Leone, although they had a large army made up of other people whom they led.

Soon, by marrying Loko, Sherbro and Temne, they became like these people. In fact a few years before 1600 a Portuguese trader said that he could not tell whether a man was a Mani or not. The people of Sierra Leone no longer thought of them as being strangers.

Seventeenth-century writers spoke of an emperor who lived in a town far from the sea called Mano (now in Liberia), and they said: 'The Bullom and Krim call

the subjects of this emperor Mendi, that is, Lords.'
This is the first time white men had heard the word
Mende.

Another European trader who lived forty miles up
the River Jong in the 1750's, and who wrote about the
people who lived in his district, said: 'As for the inland
parts, I have not learned anything except about their
nations which, by their own account, are as follows: to
the East of the Bullom lie the Temne. Next are the
Banta and then the Kono and the Tene.' The Tene are
a people in Liberia. The Banta are related to the
Temne.

The Mende, therefore, although they were not far
off, were not in Sierra Leone by 1750. But they soon
began to attack it. For instance, it is said that a Muslim
Fulah went to Mende country and became a great man
with many wives, and he was friendly with the princi-
pal men of the country. His son, taking the name
Fulah Mansa, that is 'King Fulah', became a warrior
and, about 1786, conquered the Yonni district.

Early in the nineteenth century the Mende were
called Kossoh and, although this word now is impolite,
they did not seem to mind at the time. This word in
Songhai means 'a young man'. In 1591 the large
Songhai empire, which held many different peoples,
was set upon by warriors sent across the Sahara desert
by the Sultan of Morocco. They were led by a Spaniard
called Juda Pasha who defeated the Songhai army at
the battle of Tondibi.

Many peoples, once under the Songhai, now saw
their chance to break away and become independent.

It seems that this was what the Mende did; the young men, the *kossoh*, the warriors, set out to look for a country of their own. After many years' hard fighting they found a road to the sea and settled beside the Sherbro.

But they did not stop fighting; in fact nearly everyone fought at one time or another in the nineteenth century. In the north the Susu met the Temne, Limba and Loko. In the south the Temne attacked the Sherbro, and were in turn set upon by the Mende. The Mandingo Alimami Samory, or Samodu as he is sometimes called, attacked the Koranko who had moved into Sierra Leone from the north about a hundred years before. Samory said he wanted a road to Freetown, but he was told by the government in Freetown that there could be no talk of such a road until he took his warriors back over the Niger River.

The new Colony of Sierra Leone was raided several times; rich Creoles like Sir Samuel Lewis began to say that trade was upset by all these wars, and so in 1896 the land beyond the Colony was made a Protectorate. After the rising of 1898 there was peace at last. Soon the Sierra Leone Railway was built and there was more trade than ever. But this is to look too far ahead; we must first see how the Colony began.

6. *The Slave Trade and the Founding of the Colony*

SLAVERY was not something new; for example there had been slaves in Greece and Rome before Christian times began, and there had been slaves in Africa since before the days of ancient Egypt.

The first slaves were taken from the West Coast by the Portuguese in the 1440's. The Portuguese explorers had wanted to show their royal master that they had indeed passed North Africa and so sailed beyond the lands of the white men. Also, possibly, it was felt that it would be good to teach the slaves Christianity.

So far, the merchants in Portugal had shown little interest in the voyages of discovery because they saw that not much profit could be had from them. But unluckily it happened that after 1492, when the New World was discovered, much of it was found to be too hot for a white man to work on his farm as he could do in Europe. Also, as in West Africa, there were many diseases and white men died quickly.

European merchants now saw that big profits could be made by carrying Africans across the Atlantic Ocean where they could be sold to work on the farms (or plantations, as they were called). They grew such crops as sugar, tobacco, cotton, indigo, ginger and coffee.

For the first time the slave trade now became widespread; it involved very large numbers of people.

28

Soon other nations copied the Portuguese; John Hawkins by his slave voyages between 1563 and 1568 made himself within a few years the richest man in England. Until the seventeenth century this trade was mainly in the hands of the Portuguese, but after that time the French, the Dutch and the English became the chief traders.

Happily, few slaves came from Sierra Leone. This was because Sierra Leoneans were not popular with planters in the New World and because in the eighteenth century, when slavery was at its height, Mandingoes were especially welcome as house servants. Also the Royal African Company, an English Company and the chief exporters in Sierra Leone at the time, only once brought goods worth as much as £5000 to this country in one year: yearly shipments were usually about half that. Many of these goods were exchanged for ivory, redwood or Sherbro mats; there was little left over, therefore, for buying slaves.

In 1772 there were about 15,000 slaves in England, who had been brought by their masters from the New World. In that year by a judgement in the law courts it was held that these men, in a country of free men, were also free. The name of the slave who won this very important case was James Somerset; he had been helped in his fight with money and advice given by an Englishman called Granville Sharp.

Sharp (1735–1813) was the grandson of an Archbishop of York. He had taught himself Greek and Hebrew so that he could better study early Christian writings. In 1767 he had lost a law-suit against

29

another slave owner in a similar case, when the judge had decided that a slave was still a slave even when brought to England. Sharp then set himself to fight this ruling, both by his pen and by actions in the law courts. In 1787 he founded the Society for the Abolition of Slavery; he was also one of the founders of the British and Foreign Bible Society.

James Somerset had been brought from Jamaica by an Englishman called Charles Stewart, to act as his servant. He soon found a chance to run away, but Stewart found him and put him in chains on a ship called the *Ann and Mary*, sailing for Jamaica.

By English law there is a rule that no man can be imprisoned without a proper reason being shown. A letter, called by lawyers a writ, can be sent to any gaoler asking him to produce the person concerned and to say why he is keeping him in prison. This writ is called *habeas corpus*.

Somerset was demanded and produced by this means. The question had then to be settled whether English law applied to a slave from Jamaica, or whether Jamaican law applied. English law does not and did not allow slavery; Jamaican law did. It took the lawyers many months to decide that English law applied and that Somerset could not be put in prison by Stewart, nor be a slave, nor be sent back to Jamaica.

This case was important because judges who work in law courts where English law is used decide cases largely by looking up decisions of earlier judges in similar cases; therefore any other slave who was said not to be free in England could quote the case of James Somerset.

Many of these men, though now free, were unhappy about settling in a cold, strange country. They had no money. No sugar, cotton, ginger or coffee grew on the farms there and many could find no work.

During the war of American Independence, which began in 1776 and which gained for the white American settlers their freedom from England, slaves who fought for England against the Americans were promised their liberty.

In 1783, when peace was signed with the new-formed United States of America, these slaves were taken to the West Indies, to Nova Scotia or to England— mainly to London, where they often wandered as beggars in the streets.

By 1787 there had been for several years a feeling amongst humane people in Europe, especially in England and Sweden, that slavery was wrong. The British Government soon decided to send some of these former slaves to Africa, whence either they or their parents had first come, and Sierra Leone was chosen as their home. Granville Sharp, with his own money, bought them some luxuries to take with them.

They sailed from England on 9 April 1787 and arrived in Sierra Leone on 9 May. On the 11th they saluted the Temne King Tom with thirteen guns before Captain Thompson, who had brought them out, went ashore to speak with the King and to arrange to pay him for land where the Settlers could live.

This land was on the Sierra Leone peninsula, now within the Western Area (which from 1808 to 1961 was a Crown Colony). It was only in 1896 that the peoples living inland were joined to the Colony. That part of the country was then called the Protectorate.

A settlement was made with King Tom on 14 May 1787, and on the next day the newcomers cut a clearing through the bush to the place where State House now stands, and hoisted the British flag. There a meeting was held to choose a leader, and Richard Weaver became their chief magistrate. The next Sunday the Reverend Peter Frazer, their chaplain, said divine service. But there was some doubt over the agreement with King Tom and the settlement had to be confirmed by his superior Naimbana on 12 July 1787.

Then plots of land of one acre each were allotted and a site chosen for the town. This was by Kru Bay and they called it Granville Town after Granville Sharp. A store house was begun, but building materials were scarce—especially oyster shells from which lime for the cement was made. The Temne were suspicious and another land agreement had to be made with Naimbana on 22 August 1788. By now the land had been cleared, but only a few houses had been built and there was trouble in getting enough drinking water.

In November 1789, in revenge for an attack by white slavers upon his town which lay half a mile from where the Settlers lived, a Temne chief attacked Granville Town and destroyed it. Messrs John and Alexander Anderson, a slaving company on Bunce Island, gave shelter to the survivors.

32

An ivory salt-cellar, carved for the Portuguese about 1550, in the same style as the nomoli *or soap-stone figures found in the earth on Sherbro island, and the nearby mainland.*

(a) Suri, King of Sherbro, *1758. Drawing by Nicholas Owen, reproduced from his Jou*

(b) *Governor Rowe on 31 October 1877 addresses chiefs from Mellacourie and S Bullom districts during British–French rivalry in the Northern Ri Lawson interprets—see p*

3　(*a*) The '*Black Joke*' *from Freetown captures the Spanish slaver* '*Almirante*' *with 460 slaves on board, 1829.*

(*b*) *View of Freetown in the 1870's.*

4   *Reclining figure carved in soap-stone, an outstanding example of* nomoli *of which many hundreds have been found in Sierra Leone.*

Early in 1791 Mr Falconbridge, an Englishman, was sent out as Chief Agent for the newly formed St George's Bay Company. This had been set up especially to help the surviving Settlers, who now numbered less than a hundred; once again Granville Sharp was a leading figure in the work. Later in 1791 the Company's name was changed to the Sierra Leone Company.

The new settlement took shelter in seventeen houses deserted by their owners because of 'devils' supposed to be in them. They were near the modern Cline Town; once again the settlement was named Granville Town.

In 1791 silver and copper coins were introduced (there were silver dollars, half dollars, twenty and ten cent pieces; there were copper one cent and penny pieces). On one side was a picture of a lion and on the other two hands clasped to signify friendship between the black and white people here. The coinage as we know it today was introduced in 1964.

The Nova Scotians arrived in March 1792 led by Lieutenant John Clarkson, an English sailor. These people had been unhappy in Nova Scotia because the land they had been given after the American War was so poor. One of them, Thomas Peters, went to London and asked the government to send them somewhere else. They now settled near the first Granville Town of 1787 and they named it Freetown. The Company made Clarkson their Govenor.

But the Nova Scotians were not happy here either; disappointed by promises never carried out in Nova

33

Scotia, they now found themselves neglected in Sierra Leone. They had few houses, for instance, so that they lived in tents made from the sails of ships. Instead of the twenty acres of land they had been promised they received only four, and some of these, they said, were too poor and too hilly to grow anything in the first year. Food therefore became scarce. There was in fact a reason behind this smaller grant of land: to give everyone twenty acres would have made the settlement too widely spread to defend.

At the end of September 1792 yet another land settlement was made with the Temne, and this seems to have reduced the Settlers' property still further; the area once granted by Naimbana had now been made very much smaller.

However, by the end of 1793 Freetown was greatly improved. The Governor's house stood near Water Street, not upon the hill now called Tower Hill. The settlement along the seashore was protected by some twenty-seven large guns, some round the Governor's house, some on a platform in the middle of the town and some at the west end where people landed from ships. A court house had been finished, built in the style of the West Indies with galleries all round, like many a house in Freetown today. Each house had a small garden. In the church, which had seats for 1000 people, the schoolmaster acted as minister.

There were houses too for the doctors, engineers, cashiers, store-keeper, commercial agent, and a few others. There were two hospitals, a boys' school and a girls' school, but the other buildings were still very poor. Labourers were paid 1s. 6d. a day and worked from sunrise until sunset with a break from 8 a.m. to 9 a.m. and another from 11 a.m. to 2 p.m. Eggs cost 2s. a dozen, which was dear for those days.

Then disaster struck again. On 28 September 1794, because France and England were at war, the French attacked Freetown. They sailed into the river with their men dressed like English sailors and with their ships showing the English flag. Freetown was taken by surprise and everyone fled to the shelter of the woods which lay about three miles out of town. The enemy stayed until 13 October, killing all the live-stock and leaving the Settlers with almost no food; they also carried off all the coins they could.

The Nova Scotians, already disappointed, now be-came angry at their treatment and turned rebellious. In 1799 they began to say that they had as good a right to make laws for the settlement as had the Com-pany's Governor; that the Sierra Leone Company had no legal power to enforce its rights and that they could expel all Europeans from the settlement if they so wished.

Then the Maroons sailed into Freetown at the end of 1800. Long in rebellion in Jamaica them-selves, they had at last made peace and they helped to put down the new rebellion. Therefore, instead of being settled as was first planned on the Banana Islands and the Bullom shore, they were housed in Freetown where they had been so useful.

The name Maroon has an ancient history. It comes from the Spanish *Cimarron* which means 'one who goes to live in the mountains'. In 1509 the Spaniards had taken Jamaica, but the British captured it in 1655. By 1660 the last Spaniards were driven out and their slaves, many of them Ashantis from the Gold Coast, went to live in the mountains, until persuaded to sail to Nova Scotia. From there they came to Freetown.

From these three, the original Settlers, the Nova Scotians and Maroons, today's Creoles are descended. But Freetown's Creole population was later increased by Liberated Africans taken from slave ships after 1807. Local peoples coming to town to copy their way of life also became part of Creoledom.

In Sierra Leone the Sierra Leone Company, finding their control weakened, had asked the government in England for a charter, which they were given in 1799. This made the settlement an independent colony and gave the Company's directors some right to make laws like those in England, to appoint a governor and council and to set up a mayor's court to hear law cases about private people. More serious cases, such as murder, were heard by the council.

Before this new power could be fully tried out a further disaster overtook the Colony. On 18 November 1801 the Temne attacked, led by Nova Scotian rebels. In March 1802 a truce was made, but the Temne were still worried about the Colony's growing strength and they attacked again in April. The Company's directors now asked the British Government for help—first for money, which was refused, and then for soldiers. After some years' talk it was agreed

36

that the Colony should accept government from England.

On 1 January 1808 the Company surrendered the Colony to the British Government, and the Sierra Leone settlement thereby became a British Crown Colony. That is to say the British Government in England, and not the Company, now appointed the governor, and the people living in the Colony became British subjects; the governor was responsible in the end to the British Government.

A colony in British territory differs from a protectorate in that a protectorate is governed by the British Government, but the people there are not fully British subjects; nor do they come under the same laws as do colonial subjects.

7. Nineteenth-century Freetown

IMAGINE a boy born in Freetown about 1830. In those days there were no trains, cars or motor buses, and everyone either walked or rode a horse to work. Life altogether was very different from life today. The boy might have been one of the first fee-paying students at the Church Missionary Society or the Wesleyan Missionary Society, which together ran most of the schools. (There were, of course, Muslim classes too; for example, the Yoruba, or *Akus*, had a mosque near Fourah Bay.) As he went to school in the morning he would meet friends on the way; the boys wearing a shirt and trousers and the girls a long dress with a short jacket of blue and white check. Girls were taught reading, writing and needle-work; boys learnt reading, writing, arithmetic, English grammar and geography; all received religious instruction.

There were fourteen townships round Freetown. The largest were Regent, Kissy and Gloucester. Regent had five stone houses and was proud of them; it was also a famous centre of learning. On the way to Regent, at Leicester, there was a hospital for sick people who had been rescued from slavery by the naval squadron in Freetown. Such freed slaves were called Recaptives or Liberated Africans and they soon came to outnumber the Settlers.

Farther on the boy might see other Recaptives building their houses; they worked four days a week on them, and for a year the men received free food until they could gather enough from their farms to support themselves. They could not build less than twelve feet away from the road, and the road itself had to be forty feet wide. At the back of the houses in these towns each family had to clear a space of another forty feet, although they were not allowed to cut down big trees without special permission from the Super-intendent in charge of that town.

Besides the three larger towns there were others: Leopold,* Charlotte, Bathurst, Wilberforce, Kent, Waterloo, Hastings, York, Wellington, Cape Shilling and Leicester. Few of them had more than thirty houses and these were only small mud huts, although there were two stone buildings usually found in each:

* Leopold is now part of Charlotte.

the church and the house for the Superintendent. An important new road had just been opened between Kent and Waterloo.

In Freetown there were about seventy-five houses made of stone and brick and about two hundred others made of wood. Retired soldiers and labourers lived in small huts in the city. All the streets were much wider than is usual today. They had been carefully planned so that many of them ran inland from Water Street, which lay along the sea-shore; others crossed these at right angles. They were eighty feet wide, but Water Street was twice that width. It is possible to find these wide streets today. There were two markets, one especially for rice and fruit and the other for meat and fish.

Let us suppose the boy's father was renting a house in town. It might cost him as much as £300 a year, but, because he was probably a Recaptive, government would have given him land in the town; this was a standard plot, forty-eight feet in width and seventy-six feet in depth. The population was a little over 13,000 with less than one hundred white people, including two or three wives. The original Settlers from England had almost died out, although twenty or thirty remained.

In the south part of the town lived the Nova Scotians. The Maroons were in the north-west and they were well known for their hard work—nearly all the houses in their district were built of stone and wood. The Recaptives, too, often lived in groups; the Kissi amongst them founded Kissi town. The wars in

39

the Congo provided a steady stream of Recaptives in the 1820's. They first settled in the hills to the south of the town, but they soon moved down to the sea—hence Congo town today.

The Temne traded with the settlement, especially in lime from oyster shells and in timber. The Sherbro brought dried fish which they often caught near the Banana Islands.

Near Lungi lived a powerful Susu warrior called Ala Dala Mahamoud. Well educated, he spoke perfect English and he was very rich; he grew coffee, rice, tobacco and cotton for sale. In a Bullom town in Susan's Bay lived a headman called King George; he had been taken as a young man to England by a ship's captain.

Perhaps the boy might want to go and see his friends in Gloucester. His father might well tell him the story of the wild elephant and her baby which had frightened so many people in that town some years before. No doubt his father would keep a pony, which would cost about £12 to buy, for his own journeys. In the evenings, when he was not busy, he would probably dress up and ride out towards Kissy race-course. All the important people in Freetown thought it fashionable to be seen there in the evening. There were few of the flies which today make Freetown dangerous for horses, and races were held at Kissy every year in December. On his way back perhaps his father would stop at one of the many public billiard rooms for a game or a chat with his friends. At eight in the evening a gun was fired to call people to their houses where they talked or slept until five when a second gun told them that another day's work had begun.

There was no regular mail-boat service and the arrival of a ship was a noteworthy event. When a vessel was seen from Signal Hill a flag was hoisted; red if the ship came from the north and white if from the south. The soldiers at the same time fired a gun. By the time the boy had left school, there was a simple lighthouse at the end of the peninsula near the rocks so dangerous to shipping. Mr Eliot, a Nova Scotian who had a farm there, had made it by placing a lantern on top of a pole. The lighthouse we know today was not built until 1850.

The time would come for the boy, now a young man, to go into business. Perhaps he would trade in Freetown, or maybe he would go to try his fortune in Bonthe, where so many Creoles were at that time. Wherever he went, no doubt he would shake his head sadly in the company of his friends and complain, as they all did, that wars in the interior would ruin him. They did not ruin him, but a merchant must grumble about something when times are bad.

Later, for a change, he worried about what the French were doing to the north of the Colony, round the Mellacourie and Scarcies rivers where so many of his friends traded. They told him that France was going to take this profitable market from them.

Meanwhile Freetown grew bigger and the colourful houses, painted blue, grey, yellow, red, green and many other shades, became more numerous. Near King Tom point was the Wesleyan College; the Colonial

Hospital had been built at King Jimmy and near it the Police Court. A little further up-river Charles Heddle, the rich merchant, was just finishing building his huge store-house with bricks brought from Marseilles in France. He was one of the first to export ground-nuts and palm-kernels.

The church of St George which had been newly built when the young man was small became a cathedral in 1852. The Church Missionary Society College, which was founded in 1827, had now been moved to a big new red building near Fourah Bay. In 1876 it was to start preparing people for the degrees of the University of Durham, England.

In the streets and clubs and private houses the boy would meet many famous Sierra Leoneans. As he became older and more respected they would stop and speak to him. There was for instance Dr Robert Smith, born in Freetown about the same time as the boy himself. From 1866 to 1885 Dr Smith was Assistant Colonial Surgeon; he always dressed very elegantly, often with a white waistcoat and a flower in his buttonhole. There was Thomas George Lawson, from modern Togo, who had been sent to school in Freetown. In 1852 he became Government Interpreter; a fervent Christian, he was also lay pastor in the Church of God in Circular Road. There was Melvin Stuart who had come from his home in the Bahamas. In 1878 he became head of the Customs Office. With his black beard and tail-coat he looked very imposing, but he was popular with his clerks in the Customs. There was the young white-turbanned Mohammed Sanusi, a

Muslim who in 1872 became Arabic Writer to the Government. He held this post until 1901.

In 1871 the West Indian scholar, Dr E. W. Blyden, had arrived and had begun to talk knowingly of exploring the Western Sudan. In 1872 he was sent to Falaba to try and find the source of the River Niger, still unknown to Europeans. When he returned he became Agent to the Interior, and in 1873 he went on a long journey to Timbo by way of Kambia on the River Scarcies. He returned in ill-health, so he resigned. He advised everyone not to send boys overseas for their education, and he talked always of the need for starting a West African University. He urged, too, the formation of a City Council so that Sierra Leoneans could have more responsibility, but this did not happen until 1893. The first elected Mayor (1895) was Mr Samuel Lewis. Born in 1843, Lewis was a leading member of the Legislative Council for over twenty years. Three times Mayor of Freetown, he was the first African to receive a knighthood.

Another famous man was J. Hastings Spaine, educated at the Grammar School, who in 1882 was transferred from a post in the Secretariat to become Colonial Postmaster. He was building himself a fine new house in Gloucester Street. There was the very tall man with a beard, Daniel Carrol. He had been one of the first pupils at the Grammar School, founded in 1845. In 1857 he began his legal career when he became Clerk to the Master of the Court; in 1882 he was appointed Master and Registrar.

There was Aaron Sibthorpe, the historian. He too

entered the Grammar School—in the same year as Dr Robert Smith. Sibthorpe became a teacher in 1865 at Christ Church, Pademba Road—then almost on the edge of the town. In 1868 he published his *History* and *Geography* but he was not the first Sierra Leonean to publish a book: Dr James Africanus Horton, a Surgeon-Major in the West India Regiment stationed in Freetown, had already written a series of books on tropical medicine. Finally, there was T. J. Sawyerr, born in Waterloo, who established Freetown's first bookshop in 1856.

By the time the boy we imagined was getting old, his sons would run the business. There would be plenty for them to do because the town's population was over 30,000; more than twice what it had been when their father was a child.

In 1893 Freetown was made a municipal city, a sign of its growing importance. New ways lay ahead: there was talk of building a railway; a regular steam-ship service had been running between Sierra Leone and England ever since 1852—the African Steamship Company, later to become Elder Dempster's. That was the year after direct taxation had been laid on houses and land—and our boy of the 1830's, now over 60, would no doubt remember how his father had grumbled at it. He would remember when the tax was taken off again in 1872 and higher Customs Duties laid instead on spirits, tobacco and gunpowder.

In 1875 he would have been one of the three hundred shopkeepers who signed a petition asking that the land between the River Scarcies and Sherbro should be annexed to the Colony so that traders there, too, should be made to pay a fair share of taxation. The British Government had refused.

He would remember the fighting in Ribbi country in 1887 and how it had overflowed into the Colony. And now he was told that there was talk of making a Protectorate; but the boy we have imagined could not live for ever, even amongst his memories, so perhaps he might have died before the Protectorate came into being on 31 August 1896.

8. *Government in Sierra Leone*

WE HAVE seen how, when Sierra Leone became a Crown Colony in 1808, the Governor was appointed from England. His Council, which ruled only a small area round Freetown, he chose himself; their duty was to give advice. But because they were mainly senior officials like the Colonial Secretary and because, in the end, the Governor could always ignore their advice and make laws by himself, the Council did not represent the people of the Colony.

In 1799 Freetown had become a municipality with a Mayor and Aldermen, but during the next century the powers given to the Town Council passed slowly to officials in government, and soon a municipal office was only a title with no duties. In 1851 Thomas Macfoy, the Harbour Master, had been appointed Mayor by the Governor's Council, but after that there was no other Mayor until 1894.

45

By 1850 there were many rich merchants in Freetown, and when a House and Land Tax (6*d*. an acre on land) was begun in 1851 they felt that if they were to be taxed by government they should be represented in government too. So, on 27 May 1863, under Governor Blackall, a new constitution was set up, dividing the Governor's Council into separate Executive and Legislative Councils.

The Executive Council was to give advice and the Legislative Council to make laws, or legislate. Although most of the members of these Councils were government officers, the Governor had been asked by the British Government to appoint four private citizens to the Legislative Council. Two of those he appointed were, in fact, government officials, but the third was Charles Heddle, son of a Scots doctor and an African mother from Senegal. His business, mainly in oil-palm produce, was so large that his advice was welcome.

Blackall now asked the richer merchants of Freetown to name the fourth member of the Legislative Council. They elected John Ezzidio, a Recaptive, originally from Nigeria, who thus became the first African to sit on this Council in Sierra Leone. He attended regularly until just before his death in 1872.

Ezzidio had landed in Freetown in October 1827 and he had been apprenticed, like many other Recaptives, to a shopkeeper, a Frenchman who had in fact named him Isadore. He quickly made money, especially after the Reverend Thomas Dove, Superintendent of the Wesleyan Mission, took him to England in 1842 and introduced him to wholesale firms there. He could now trade on equal terms with his European rivals in Freetown, who up to that time had been the only people to deal regularly with wholesale firms in England. He employed only Africans in his business, and one of his clerks, W. T. Dove, later became a rich merchant and founded a family well known today in Sierra Leone and Ghana.

In 1841 Ezzidio had built a large three-storied house opposite St George's Church, soon to become the Cathedral. He bought the land from the Jarretts, a Maroon family who had been granted it when they arrived about 1800. Ezzidio's neighbour was William Henry Pratt, also originally from Nigeria; he too built a large house and shop on the same side of the road, but facing Oxford Street.

Later Governors did not approve of Blackall's method of having members elected to the Council and from 1872 until 1924 Governors chose Ezzidio's successors for themselves.

The Protectorate, established in 1896, was administered separately from the Colony until 1924. But in that year a new constitution took the important step of making the Colony and Protectorate jointly responsible for the government of the whole country. Amongst other things, it provided for three Chiefs from the Protectorate to sit in the Legislative Council. But the eleven official members (that is, those who sat there because they held some public office such as that of Attorney General) were in the majority and only three of the ten unofficial members were elected. All three

47

came from the Colony. By 1947 one Paramount Chief had been nominated to the Executive Council.

In 1951 another change was made when unofficial elected majorities were set up in both the Legislative and Executive Councils; the first elections for the Legislative Council were held in October 1951. From these members of the Legislative Council the Governor chose six to sit on the Executive Council.

In April 1953 these six, all from the Sierra Leone People's Party which had a majority in the Legislative Council, assumed the name and duties of Ministers. In the following year Dr Margai (1895–1964), leader of the S.L.P.P., received the title of Chief Minister.

In 1957 elections were held for a House of Representatives to replace the old Legislative Council. This House included: a Speaker, four of the seven official members of the old Legislative Council who were also members of the Executive Council, fourteen elected members from the Colony and twenty-five from the Protectorate, together with one Paramount Chief from each of the twelve District Councils (first directly elected in 1956) and two nominated unofficial members representing special interests.

Elections, as we have seen, had been known in the Colony since 1924, but in the Protectorate in 1956 direct elections were quite new; in 1951 prospective candidates had been chosen by the District Councils.

In May 1960 it was decided that the Governor should no longer take the Chair at Executive Council meetings, which should now be occupied by the Premier, and that the title of Premier should be changed to that of Prime Minister. At the same time the Executive Council became the Cabinet.

Cabinet Government was introduced on 9 July 1960 when Sir Milton Margai, knighted in 1959, became the country's first Prime Minister.

In May 1960, it was decided that Sierra Leone should become an independent member of the Commonwealth on 27 April 1961.

9. *Christianity and Islam*

CHRISTIANITY entered Sierra Leone from the west, brought by the Catholic Portuguese. One of the first priests to live in Sierra Leone was Father Barreira. After Barreira, other Catholic priests settled on Tombo Island in the Sierra Leone River and at Port Loko. They stayed probably until 1670. Then there was no Christian mission here until about 1715 when Signor Joseph came. He was the first missionary of African descent in Sierra Leone.

He had lived in North America; possibly he was born there, and he went to school in England. Later he visited Portugal and became a priest. One story says that he set up his mission in what we now call Granville Town but, falling out with his neighbours who were mostly European pirates—one captured his ship and held it to ransom—he moved to what we now call Kissy. After a long life he was buried on the Banana Islands.

The next missionaries to arrive were those who came

to minister to the Settlers; they were Protestants. The Church Missionary Society appointed a bishop in 1852; in 1859 a Catholic bishop came out with some priests but they all died of yellow fever within a few weeks. In 1864 the Catholic Mission was founded in Freetown by the Order of the Congregation of the Holy Ghost which is still here.

Meanwhile Islam had been advancing from the east, brought by the Arabs across the Sahara. They conquered Egypt between A.D. 639 and 642 and began to occupy all North Africa, especially in the eleventh and twelfth centuries. In 1042 a new Islamic movement was begun by a group of Berbers near the mouth of the Senegal River. The Berbers were one of the peoples the Arabs had found living in North Africa and the Sahara, and the group which started this new Islamic way of religious life came to be called Almoravids, which is an Arabic word. The Almoravids helped to spread the Faith to the rulers of the Sudanic empires of ancient Ghana, Mali and Songhai. In 1076 they took the capital of ancient Ghana, then under its Soninke emperors and at its most famous; though Ghana later regained its independence, it was never as powerful as it had once been.

Those once under Ghana now rose against her in her weakness and in 1240 Sundiata, ruler of the Mande state of Kangaba, whose ancestors had become Muslim about 1050, destroyed the Ghana capital. So arose the Empire of Mali which at its height, under Mansa Musa who made a pilgrimage to Mecca in 1324, reached as far west out of the Sudan as Futa Jallon.

The Songhai Empire of Gao—a town about half-way down the Niger—followed Mali as the most important state in the Sudan. It became especially great under its Emperor Askia Mohammed, who began his conquests after returning from Mecca in 1497, attacking the Mandingoes, Fulah and Hausa. But in 1591 this empire too was defeated—by the Islamic Sultan of Morocco who sent his troops across the desert from North Africa to look for gold.

These state-forming movements in the Western Sudan set many people looking for new homes, and they began to move elsewhere, or migrate, and to set up states for themselves like those they had left in the Sudan. For example, about the time of the rise of Mali certain Mande rulers imposed themselves on the Akan peoples now in modern Ghana. Also, the traditions of Hausaland (which Islam reached from Mali about the fourteenth century) and Nupe, Benin and Yorubaland tell of the arrival of conquerors about the fifteenth century. These too came from the Sudan and formed such states as Oyo and Benin in modern Nigeria.

We have seen already in chapters 3 and 5 how the sixteenth-century Mani and the seventeenth- and eighteenth-century Mende invasions of Sierra Leone, as well as Alimami Samory's attempts to reach Freetown in the nineteenth century, were also connected with the Mande peoples' history in the Sudan. So, indeed, were the movements of the Susu, Fulah and Baga peoples in chapter 1. This shows how important it is to learn about Sierra Leone's neighbours if one is to understand properly the history of this country.

Although the invaders who came before the eighteenth century were not yet Muslim, Islam soon followed in their footsteps and entered the forests of Sierra Leone. It reached the north and east parts of this country first and it is these districts which today hold the greatest number of Muslims.

Islam came in two main waves to the West Coast; the first was a slow one which lasted from the eleventh to the seventeenth centuries. During this time it was more a religion of certain classes than of whole nations. That is to say, the rulers of the West Sudan states were often Muslim; their people were not. Islam was carried further by the traders and wandering priests who travelled so widely among the peoples of West Africa; it was they who first brought it to Sierra Leone. The second movement began shortly before 1800 with the Fulah Holy War, or *Jehad* as it is called in Arabic, which began in Futa Jallon. This advance of Islam, though there are now no Holy Wars, still goes on.

There are, of course, other religions than Christianity or Islam. If you are a Christian and have read the New Testament, or if you are a Muslim and have read the Koran, you will know what a struggle both these religions had in the beginning against other beliefs in many parts of the world.

In West Africa those who are not followers of these two religions are usually called animists; that is to say they worship their own gods of the home and forest and they believe in spirits just as people did in Europe 2000 years ago.

About 1510 a writer said: 'All the villages of Sierra

Leone have a religion; they have idols and believe that these can help them in time of need.' This was not quite true. They did indeed have carvings, but these were not idols in the strict sense because they represented the idea of some spirit; it was really the spirit to which they prayed, using the idol as their messenger as it were. Today it is said that there are some 2,000,000 animists, 600,000 Muslims and 70,000 Christians in Sierra Leone.

People do not accept a new religion all at once. About 1606 Barreira, the Jesuit missionary, travelled east into Susu country. The king there had been visited already by travelling Islamic priests and he carried amulets or charms which they had sold him as a protection in time of war. No doubt he had heard beforehand of Barreira's work on the coast and he welcomed him, promising to destroy the amulets and also the figures representing his own gods which he still kept. In fact he did neither, and this shows he was trying out the two new religions at once—Islam and Christianity—to find which suited him best and whether he wanted to change from his own.

When the new religion is accepted, schools are set up and children sent to learn the Koran or Bible. Poro, Bundu and other societies begin to die out; families inherit property by Islamic or European laws and no longer by old custom. It must be remembered that it always take several hundred years for this to happen.

Islam and Christianity have brought other gifts besides religion to West Africa. As we have seen, most of the schools in early Freetown were run by Christian missions; early missionaries like Schlenker made the first studies of Sierra Leone languages, translating them into English in the nineteenth century. Hospitals and doctors have been provided; birds, beasts and trees have been identified. The Islamic calendar of twelve months based upon the movements of the moon has been brought in; Islamic methods of administration have been adopted by many Chiefs. African languages, which once had no alphabet of their own, are often written in Arabic script and new words enrich people's speech so that they are better able to say what they are thinking.

Questions

1 THE FIRST PEOPLE IN SIERRA LEONE

1 As you sit in your class-room, in which direction is north? Where is east?
2 Where is Futa Jallon?
3 Why do sailors need fresh water? Why did the Susu need salt?
4 What were the names of the peoples in Sierra Leone before 1500? When did modern Sierra Leone begin?
5 Where is Portugal?
6 Draw a picture of an elephant, a bow and arrow.

2 THE MANI INVADE SIERRA LEONE

1 What weapons did warriors of early days carry?
2 How were the Mani different from the people of Sierra Leone?
3 What food did people in Sierra Leone eat in those days?
4 What was the name of the Temne king who fought the Mani?
5 Who defeated the Mani in the end, and where?
6 Who was the first Mani king of the Loko?
7 Draw a picture of a Sherbro boat; a Portuguese boat.

3 KING TURA AND KING FARMA

1 Who was the king of the Mani? What does his name mean?
2 Give the names of three of Bai Farma's sons. What does the word *Farma* mean? What language is the word taken from? Why is this important?
3 Write a story about the Christian church service in 1605.
4 Where were the Empires of Mali and Songhai? How long ago did they exist?
5 Where did the Mani come from? (See also chapter 2.)
6 Draw a picture of the wooden church in Salvador and some of the sights to be seen nearby.

4 THE EARLY EUROPEANS

1 What is a company? Name four in Sierra Leone today. Do you know the names of any in the seventeenth century?

2 What do we mean by the seventeenth century? Give the dates of the first century A.D.
3 Draw a picture of a modern steamship.
4 What goods did early Europeans take from Sierra Leone? What is exported today?
5 How long does it take to go by sea from Freetown to England? Why did it take longer in 1600?
6 Where were the European settlements made in Sierra Leone? Draw a map showing these.
7 Where are the West Indies?

5 THE COMING OF THE MENDE

1 Where did the Mende come from?
2 When did the Europeans first hear of them?
3 What do you know about Alimami Samory?

6 THE SLAVE TRADE
AND THE FOUNDING OF THE COLONY

1 Do you know who was the most famous royal person to encourage explorers in Portugal?
2 What do we mean by the New World? Why is it so called?
3 We talk about 'Christian times'; what do the letters B.C. and A.D. stand for? What languages do they come from?
4 Can you remember what trade goods Europeans brought to Sierra Leone?
5 What kind of law do magistrates use in this country? Is any other kind of law used? If so, where?
6 What do you think people used for money before coins were brought to Sierra Leone?
7 Draw a picture of a house like that built as a court-house in Freetown and copied from those in the West Indies. How does (a) hot sunshine, and (b) cold weather make a difference to the kind of houses people build?
8 On your map find out where are: England, Sweden, Nova Scotia, France, Spain, the United States of America, the Atlantic Ocean, Jamaica.

1 If you live near the railway, can you find out when it first reached your town? What other forms of transport are, or have been, used in Sierra Leone? Give their dates as near as you can.

2 If you live in Freetown, can you find where the old streets are?

3 Do you know the name of the fly which is so dangerous to horses? Is it dangerous to other animals?

4 What has happened to the French settlements to the north of Sierra Leone?

5 What is the difference between a church and a cathedral? How many bishops are there in Sierra Leone and what are their names? (See also chapter 9.)

6 What are Customs duties? How do they affect you?

7 Why, do you think, did the government in Freetown need an Arabic Writer in 1872? (See also chapter 9.)

8 What is the name of the present Mayor of Freetown? What, do you think, are his duties?

9 What are the duties of: a Postmaster-General, a Registrar of the Law Courts?

8 GOVERNMENT IN SIERRA LEONE

1 Do you know what a Town Council does?

2 What is meant by a wholesale firm?

3 What do we mean by the constitution?

4 What is an election? How does it work in Sierra Leone? Who is your Member of Parliament?

5 What is the name of the Speaker in the House of Representatives? What are his duties?

6 What, briefly does the Cabinet do?

7 What are the duties of a Prime Minister? What is the name of the present Prime Minister?

8 Name four Ministries and give the names of their Ministers.

9 Do you know what is meant by the titles: Ministerial Secretary, Permanent Secretary?

9 CHRISTIANITY AND ISLAM

1 Why did Christianity come from the west and Islam from the east?
2 What is the difference between: an animist, a Christian, a Muslim, a Catholic and a Protestant? What is a Jesuit?
3 What is a *Jehad*; do you know any stories about one?
4 Of what religion were the first people in Sierra Leone?
5 Why do you think there are more animists than Muslims and more Muslims than Christians in Sierra Leone?

Explanation of Difficult Words

Words are explained in the sense in which they are used in this book. This does not mean that there are no other senses. The list is not a substitute for a dictionary.

account　a record or story.

account books　books where accounts or records, usually about money, are kept.

Admiral　a senior officer in command of warships.

alliance　a treaty, or agreement of friendship.

annexed　to *annexe* is to take over.

antiseptics　medicines used to kill germs.

anxiously　uneasily, full of worry.

apprenticed　to *apprentice* is to place someone in the charge of another person to learn a trade.

attend　to *attend* is to be present.

billiards　a game played with balls hit with a stick on a flat table.

cannon-balls　heavy balls fired by very large guns called cannon.

cask　a wooden drum, often for storing water.

celebrate　to *celebrate* something is to hold *a ceremony* about it.

ceremony　a public action usually to do with religion or government.

chaplain　a Christian clergyman.

charter　a written agreement by which a ruler or a state gives certain rights.

christened　to *christen* is to give someone a name during a special ceremony in the Christian church.

commercial agent　a person who acts in business, or commerce, for someone else.

Commonwealth　The Commonwealth is made up of self-governing nations who agree that the Queen of Great Britain is the Head of the Commonwealth.

company　a number of people who have made an agreement, often for trading purposes.

59

concern   to *concern* is to have to do with.

confirmed   to *confirm* is to agree to something which has already been done.

constitution   laws set up by a state for its own government and which everyone must obey. Sometimes the constitution is a written one, like that of the United States of America or of France. Britain's constitution is largely unwritten.

create   to *create* is to make.

crew   the men who *sail* (see below) or paddle a ship.

current   the running, or flowing, of water.

deserted   to *desert* is to leave alone or empty.

diary   a note-book, in which something is written every day about the events of the day.

direct taxation   a tax paid directly by anyone to the government; *indirect taxation* is a tax usually paid to someone else who then pays it to the government.

director   one who guides, or directs, the work of a company.

disappointed   to be *disappointed* is not to get what one had been promised or has hoped for.

disaster   bad fortune.

divine service   Christian worship.

elegantly   in a way people like looking at.

emperor   a man who rules over a number of peoples.

empire   groups of peoples ruled over by one ruler.

establish   to *establish* is to set up.

export   to *export* is to send goods out of a country.

fall out   to *fall out* with is to become unfriendly with.

fashionable   smart, up-to-date.

fatal   causing death.

fort   a building or group of buildings defended by guns

founding   beginning.

galleries   a gallery is a long passage, at the side of a house, with a low wall and a roof.

gaoler   a man in charge of prisoners in a prison.

graze   to *graze* is to eat grass.

harsh   unkind.

hobby   a work one loves and does in one's free time.

hoisted   to *hoist* a flag is to raise it on a flag-pole.

humane   kind.

identify   to *identify* something is to discover what it is called (perhaps in another language) and what sort of thing it is.

ignore   to *ignore* is not to listen to.

impose   to *impose* is to set up above.

imposing   commanding, important.

interloper   a person who trades without a licence.

introduce   to *introduce* is to bring in; or to bring people together.

invade   to *invade* is to enter a country as an enemy.

invented   to *invent* is to make something new.

involved   to *involve* is to *concern* (see above) or include.

Jesuit   member of a Roman Catholic religious order, the Society of Jesus, begun in 1534.

judgement   what a judge says, or decides, often called a decision.

lay pastor   a person who is in charge of a church but not a clergyman.

law-suit   a law case.

livestock   animals belonging to somebody.

luxuries   food or goods better than is usual.

magistrate   a man who is in charge of carrying out laws.

Mass   a special ceremony, or service, carried out in Roman Catholic Churches.

Mayor and Alderman.   In Freetown today the sections, or Wards, elect Councillors. These Councillors then choose from among themselves the Aldermen and the Mayor. The Mayor is the Chairman of the City Council.

measles   a bad fever which brings spots to the skin.

memoirs   a book written about what one remembers of the past.

municipal city   a city having self-government.

neglected   took no notice of.

nominated   to *nominate* is to name or to appoint.

peninsula   land almost surrounded with water, but not an island.

pretending   to *pretend* is to offer something as true, which is not.

private traders   traders not members of a public *company* (see above).

prospective candidates   those who ask to be elected.

publish   to *publish* is to make public, to print and offer for sale.

quote   to *quote* is to say or to write something already said or written by another person.

ransom   to hold to *ransom* is to steal something and make the owner pay to get it back.

reduced   to *reduce* is to make smaller.

relics   remains.

revenge   a punishment, not by the law but by another man, for a bad thing somebody has done.

rival   a person who tries to do better than someone else.

rubbish heap   a pile of anything thrown away.

ruling   a decision.

sailing   to *sail* is to spread a cloth, usually made of canvas, to catch the wind so that the wind will drive the boat or ship forwards. Steamships were *invented* (see above) only in the nineteenth century.

set upon   to *set upon* is to attack.

shipments   what is carried on a ship, often called a cargo.

signify   to *signify* is to show.

site   a place to build on.

slip-way   a floor built to run ships in and out of the water, when they are too heavy to push over the rough ground like a canoe.

source   a place where something begins or is first found.

standard   fixed by rules.

state   people living under one ruler, but not so many different groups as those in an *empire* (see above).

station   to *station* a regiment is to send it to a certain place.

subject   people ruled by another.

suspicious   suspecting other people, doubting their good-will.

title   a name showing a person's work or his importance.

tomb-stones   stones placed in a grave-yard.

truce   a peace agreement, but for a short time only.

voyage   to *voyage* is to travel by sea.

wholesale firm   a business which sells goods in large amounts to shops (retailers) which sell them again in smaller lots.

8655